THE LITTLE BOOK OF REAL ESTATE MATH

SAM MARTIN, PHD
SUE MIRANDA ROSENSTEEL

The Little Book of Real Estate Math

Published by Crimson Sparrow

Please check out Sam's blog: www.samthetutor.com/blog

Printed in the United States of America.

I want to thank my wife, Prescila, without whose patience and love I would be as nothing.
My daughter, Jamie, and son, Phil, are always my inspiration.
Without family, nothing would get done.
Dr. Sam

I wish to dedicate this book to all the students who allowed me to participate in their life journey and help make a positive difference in their lives. You have inspired me in countless ways!
Sue Miranda - Rosensteel

INTRODUCTION

Welcome to the exciting world of real estate math! All right, so maybe it's not *that* exciting. But that doesn't mean it has to be boring, either. The purpose of this book is to help with introductory math concepts. The math dealing with the various disciplines within Real Estate: Appraisal, Sales, Brokerage, Property Management, Investment, etc.

This book will benefit anyone taking classes in real estate preparing to take a high-stakes, math-oriented exam.

Thanks for sharing the path with me. It is an honor to join you in this journey. If you have questions, please ask! Sammartinphd@gmail.com

TEST-TAKING TIPS

The course or licensing exam you are about to take is called a "high stakes exam". It is an exam designed to cull the wheat from the chaff among licensed candidates. The final exam or state license exam are tests designed to change lives; so, no pressure there, eh?

Consequently, you prepared thoroughly to take this high-stakes exam. You are embarking on a military operation to pass an examination. The following items are excellent, broad test-taking tips:

- I suggest you devote 1-2 hours outside the classroom to studying for every hour you spend in class.
- Make flashcards by hand. Although they can be bought pre-made, buy some 3x5 index cards and make the flashcards manually. This accomplishes three objectives:

1) Look up keywords/phrases and write them down yourself.
2) Writing the key terms burns a path between fingers and brain.
3) Have a friend help. Let them hold the flashcards up. You answer the card. Since the answer is on the back of the card, your friend need not know the subject; they only need to see if your response matches

the back of the card. Following these three pointers helps anchor the material in your short-term memory and assists the transfer of knowledge into long-term memory.

- If your text has end-of-chapter quizzes, take them. Try to time yourself and take between 30 to 60 seconds per question. Grade yourself when you've completed the quizzes. You might have a low score at first. With practice, this should improve. **TIME MANAGEMENT IS CRITICAL!**
- Create a spreadsheet (Excel, paper and pen, etc.) for **ALL** critical deadlines and periods. In one column, the name of the law; next, the period/deadline; comments; and the page number where it is located in your text. This is a convenient way to find this critical information without rummaging through a five hundred page textbook. (Examples: how quickly must earnest money be deposited? The right of rescission period for failing to provide a disclosure form promptly; etc.)
- Some texts have complete tests at the back of the book or on a media file that accompany the text. **Do NOT use that test until you are ready to take the final/state exam.** These tests can be invaluable assessment tools, and you do not want to waste them. At this stage, you need questions —many questions. Once you take one of those tests, you are done. When you retake these practice tests, you memorize the answers. The test is no longer "fresh." Set up a small space in an office-like setting. Prepare yourself for testing by obtaining a monitor, keyboard, mouse, mousepad, pencil, calculator, and a piece of scratch paper. Emulate the test center as best you can. Take the test (again, between 30 and 60 seconds per question), and score yourself. It would be best if you tried to get a score of 90% or more. Because the state licensing exam is stressful ("high stakes"), you can deduct about 10 to 15 percentage

points off your score and come close to what you might get on the state exam. Therefore, if you get a score of 80% on the practice test, you might get a 65% or 70% on the state exam. These scores are not incredibly comforting since a passing score is typically between 75% and 80%. But, if you get a score of 90% or more, your state exam score should translate from 75% to 80%. A happier ending, indeed!

The previous tips deal with *broad topics* about test-taking. The following tips deal with *specifics* in the proper methods of test-taking.

- As soon as you sit down to take the test, write down everything you fear you might forget on your piece of scratch paper. Every troublesome acronym, number, mnemonic device: all fair game for that piece of paper. That way, an hour into the test, you won't wrack your mind: "How many square feet in an acre?" No Problem! You already wrote that information down. In the days leading up to the exam, practice writing key elements down on scratch paper. On the day of the exam, the information will flow more readily on the scratch paper your proctor gives you. Leave nothing to chance.
- When reading a long problem (especially a math problem), *read the question first*. Ask yourself, **WHAT** is the question asking? **THEN,** read the choices. This will tell you the format of the answer: is it word, dollars, dollars/square foot, or yards? And, **FINALLY**, with great attention to detail, read the **ENTIRE** problem. Reading the problem after reading the question allows the test-taker to read the problem with knowledge aforethought. With knowledge comes power. I am asking you to buy a murder mystery, open it to the last page, and spoil the mystery. A terrible idea for a murder mystery, but the ideal solution for a real estate mathematical word problem. For example:

1. George owns a brick ranch in Joe Blow Estates. George's attorney is Samantha, who is a Leo. George sold his house through ABC Realty. The listing agent is Samantha. A 5% commission was charged; the house sold for $200,000. The assessed value is $66,600. The temperature is 72 degrees. What amount of commission was paid by George?
2. $10,000
3. $15,000
4. $20,000
5. $40,000

What should we read first? *The question!* "What amount of commission was paid by George?"

Then what do we read? *The answers!* The format of the answers is dollars. And, finally, what should we read? *The problem!* We know we don't care about George's name or Samantha's name, or that the house is brick. We care about two items: sales price and commission percentage. Even if you have never calculated a commission before, this process still demonstrates the power of asking the question first!

- Make sure your calculator is: (a) allowed by the testing service, (b) one you are adept at using, and (c) sufficient for the task at hand. You don't want to buy a fancy calculator the day before the test and struggle to figure out how to use it while trying to pass a high-stakes exam. You also don't want to get a calculator that does more, or less, than what you need. Ease of use might only come after many days of practice. For many real estate exams, a simple $10 calculator is sufficient. Get fancy later in your career when you need a more sophisticated device.
- Individuals about to take licensing exams (or finals) will find this book beneficial, as there are numerous practice math problems *AND* practical advice on how to take standardized tests. Please remember, this book is a general

overview. Specific laws and assessment rates vary from jurisdiction to jurisdiction.

The following page will take you back to your early days in school, using fractions, decimals, and percentages, relearning math... fun!

Don't forget: the answers are all worked out for you at the end of this book. Let's get started!

PRACTICE PROBLEMS

1. CHANGE THE FOLLOWING PERCENTAGES TO A DECIMAL.

12.5%		34%		5%	
125%		34.5%		5.5%	
14%		.25%		33%	
140%		.45%		25%	
9%		.34%		7.5%	
19.5%		.034%		.09%	

2. CHANGE THE FOLLOWING DECIMALS TO A PERCENTAGE.

.15		.02		.29	
.025		.257		›	
.0976		.0078		.154	
.2		.26		.788	

3. CHANGE THE FOLLOWING FRACTIONS TO DECIMALS.

$\frac{7}{8}$ $\frac{15}{16}$ $\frac{5}{8}$ $\frac{3}{4}$

4. CHANGE THE FOLLOWING PERCENTAGES TO A DECIMAL.

$7\frac{7}{8}\%$ $9\frac{15}{16}\%$ $14\frac{5}{8}\%$ $7\frac{3}{4}\%$

NOTES

Please feel free to write any notes in this section of the book.

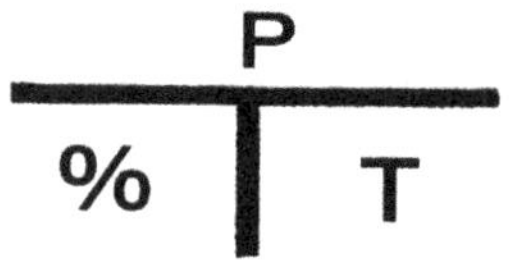

THE "T"

One of the essential tools you will need for real estate math is the "T." The "T" constitutes roughly 75% of the math in a real estate course.

The Part divided by the Total equals the Percent (see Example 1). The Part divided by the Percent is equal to the Total (see Example 2). The Percent times the Total equals the Part (see Example 3). In other words, if you go up and down, you divide. If you go sideways, you multiply.

IMPORTANT ADVICE FOR READING MATH-RELATED WORD PROBLEMS

Words associated with the "T": Sometimes, when you divide, you might see the words "is" or "per." When you multiply, you might see the words **FOR**, **BY**, or **OF**.

Example 1 = % (is or per)

Example 2 = T (is or per)

Example 3 % x T = P (for, by, of)

For example, let's look at an easy problem using a fraction. Sam owns a total of 4 houses (T). He sells 3, a part (P) of Sam's total. What percentage (%) of his homes did he sell?

= (75) % (is or per)

Or, 3 IS 75% **OF** the total 4.

Let's re-word the problem: Sam owns four houses (T) but sells 3 (P). What percentage has he sold? In other words, 3 (P) is what percentage of the total, 4 (T)?

= (75)% (is or per)

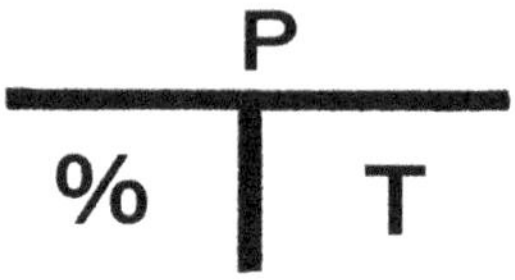

The “T” can be applied to roughly 70% of the math you will use in an introductory real estate course. We will be exploring the “T” in greater detail as we work our way through this book.

NOTES

Please feel free to write any notes in this section of the book.

"PART-PERCENT-TOTAL"

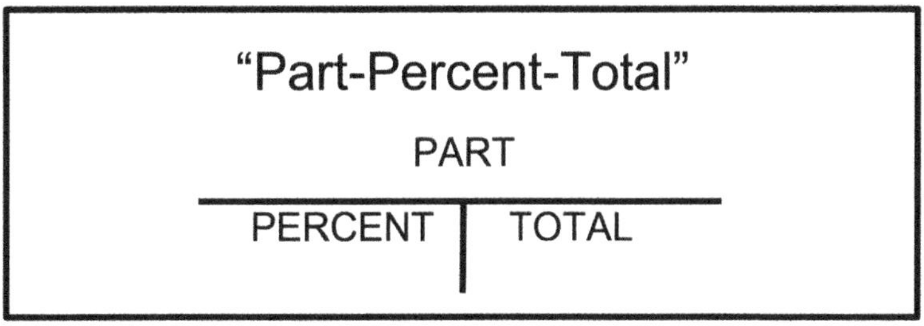

1. $8000 is equal to 8% of the price paid for a dwelling. What was the fee paid?

2. What part of an acre is 14,157 square feet?

56.0%C. 32.5%
3.0%D. 5.9%

3. A farmer wants to sell part of his land but has promised to give 1/3 to his son and 5/16 to his daughter. What part of the land is still available for sale?

56%C. 35.4%
64.5%D. 72%

4. $657.49 is 8 ½ (8.5) % of what amount?

$4,066C. $7,735
$4,260D. $4,345

NOTES

Please feel free to write any notes in this section of the book.

SEPARATE CALCULATIONS

****There may be an occasion when you must perform two separate calculations to answer a single question.****

5. A property has sold for $85,250. The sponsored licensee is to receive 12½ % of the total 7% commission. How much will the sponsored licensee receive?

6. A seller lists his home with the local broker for $78,375. The broker's commission is 7¼ %. How much will the broker earn if the property sells for 98% of the listed price?

A. $6,875.00
B.. $6,354.43
C. $4,389.63
D. $5,568.54

7. Jan, a broker, has listed a brick ranch for $55,900. When the property sells, she is entitled to 40% of the full commission collected by her sponsoring broker. If the total commission is 6¾ %, and the home sells for the listed price, what will Jan be paid?

8. 22 Twain Lane is a cooperative building. Ralph, a new member of the co-op family, has 57 shares of the 500 available shares in the corporation. What percentage of the total shares are Ralph's?

A. 114%
B. 12%
C. 11.4%
D. 4%

9.You will complete 140 questions on the state sales examination. If you must achieve a minimum score of 75%, what is the maximum number of questions you can miss and still pass the examination?

A.77
B. 70
C. 35
D. 30

NOTES

Please feel free to write any notes in this section of the book.

INSURANCE

10. A home valued at $140,000 will be insured to a maximum of 80% of the market value. If the premium rate is $3.68 per $1000, what is the first year's premium?

AD VALOREM (REAL ESTATE) TAXES

AD VALOREM (Real Estate) TAXES

Ad Valorem Tax Calculation

MARKET VALUE
X
ASSESSMENT RATE
ASSESSED VALUE
X
EQUALIZATION FACTOR (if applicable)
EQUALIZED ASSESSED VALUE
X
TAX RATE
TAX BILL

(WHERE THE TAX RATE = AMT OF TAX PER UNIT OF ASSESSED VALUE.)

11. Charlie received his tax bill and was concerned by the method of calculation used by the county. Upon calling the Assessor's office,

Charlie learned his tax rate is based upon a figure of $2.50 per $100 of assessed value. The market value for the house was estimated to be $106,500. An equalization factor of 1.35 was used. The assessed value was 25% of the market value. What were Charlie's taxes?

12. If a property is assessed at $83,625 and the tax rate is set at $3.78 per $100 of assessed value, how much will the tax be?

13. The real value of a property is $75,000. It is assessed at 40% of its actual value. It is taxed at 45 mills on the assessed value. What are the taxes?

14. Patti received her ad valorem tax bill of $242.50. The tax rate for the year is $25.00 per thousand. The district assesses real property at 20% of market value. What is the approximate value of Patti's property?

NOTES

Please feel free to write any notes in this section of the book.

TRANSFER TAX

- **Transfer tax can vary from state to state.**
- **Don't forget to exclude any assumed mortgage or personal property.**
- **Do they ask, "to the State"?**
- **Do they ask, "to the County?"**
- **Do they ask, "How much tax must be paid?"**

15. Dave sold his home for $49,500. How much transfer tax is paid to the state?

16. Jack sold his home for $63,495. The buyer assumed the mortgage of $22,875. How much must Jack pay in transfer tax to the state?

NOTES

Please feel free to write any notes in this section of the book.

SELLER'S NET

When calculating the sales price while knowing the net to the seller, the formula is the same:

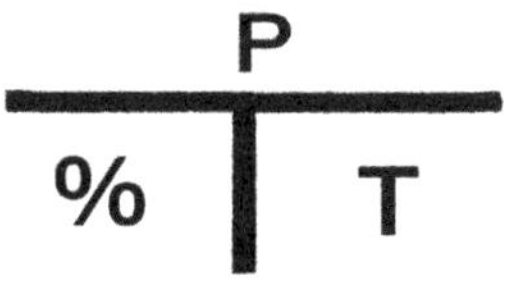

17. Jim wants to net $75,000 from the sale of his home. Your commission schedule requires you to charge 7% for your service. What is the minimum amount of money a buyer can pay to cover your commission and the seller's net?

(Hint: $75,000 is NOT the total. It is the part Jim wants to net? $75,000 is 93% of the total (100%-7%), which in this case, is the sales price.)

18. Broker Davis is listing Smith's house this afternoon. Davis knows the market value for this type of house is approximately $74,000. However, Smith wants to net $71,100 from the sale. If Davis charges a

6% commission, what would be the minimum amount Smith can profit from selling his house?

19. Jay listed his property with Dave. Jay wants to net $58,900 while Dave charges a 6½% commission on all his sales. Assuming the closing fees will total $2500, what is the minimum amount of money a buyer can offer to cover all expenses?

NOTES

Please feel free to write any notes in this section of the book.

PROFIT & LOSS

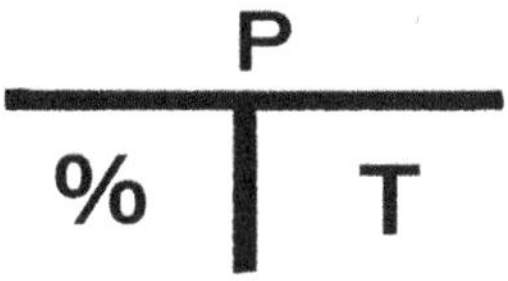

20. If you bought two lots for $4,000 each, made three lots out of this parcel, and sold the lots for $3,000 each, how much did you make? What is the percent profit?

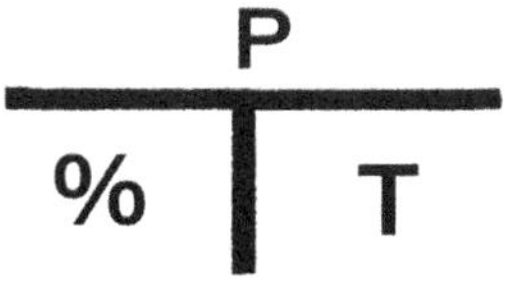

In profit and loss problems, when we're looking for the purchase price, we will use:

21. John sold a property for $125,000, a 25% profit over what he paid for the property five years ago. How much did John pay for the property?

22. In the above problem, the $125,000 represented a 25% loss. How much did John pay for the property?

NOTES

Please feel free to write any notes in this section of the book.

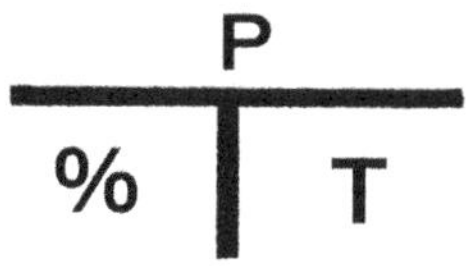

DEPRECIATION

23. An owner depreciates his apartment complex $15,000 in one year. He has:

1. Decreased the market value by $15,000
2. Increased the market value by $15,000
3. Increased the book value by $15,000
4. Decreased the book value by $15,000

24. John purchased a tract of land at a cost of $85,000 and built an office building costing $425,000. It is estimated that the improvements will have an economic life of 40 years. If John's accountant uses the straight-line method of depreciation, at the end of 12 years, what will the book value of this property be?

(Assuming land value has not increased over 12 years.)

ACREAGE CALCULATIONS

25. How many acres are in the southern half of the northern half of the southwest quarter of Section 9?

26. What is the area of:

The N ½ of the SE ¼; and the SW ¼ of NE ¼ of Section 21, all in Township 14 North, Range 3 East of the Third Principal Meridian.

NOTES

Please feel free to write any notes in this section of the book.

AREA CALCULATIONS

27. Given a room is 26 feet by 34 feet, what is the area:

  1. in square feet?

  1. in square yards?

28. Draw the parcel that is described: Beginning at a point 200 feet east of the old oak tree, then 50 feet easterly; then 100 feet southerly; then 100' westerly; then 50 feet northerly; then 50 feet easterly; then 50 feet northerly, back to the point of beginning.

  1. What is the area of this parcel?
  2. If land has a value of $5.00/SF, what is the value of this parcel?
  3. What percentage of an acre is this parcel?
  4. How much fence is required for this parcel?
  5. If fence costs $6/FT, how much would it cost to fence this parcel?

29. Calculate the value of a parcel of land with a frontage of 64 feet and a depth of 205 feet. The land is selling for $6.08 per square foot.

30. A purchaser has decided to insulate his dwelling by covering the attic floor, measuring 24 feet by 32 feet, with insulation. Each bag of insulation will cover 36 square feet. How many bags will be required?

NOTES

Please feel free to write any notes in this section of the book.

AREAS AND VOLUMES

31. What is the area of the diagram in square feet?

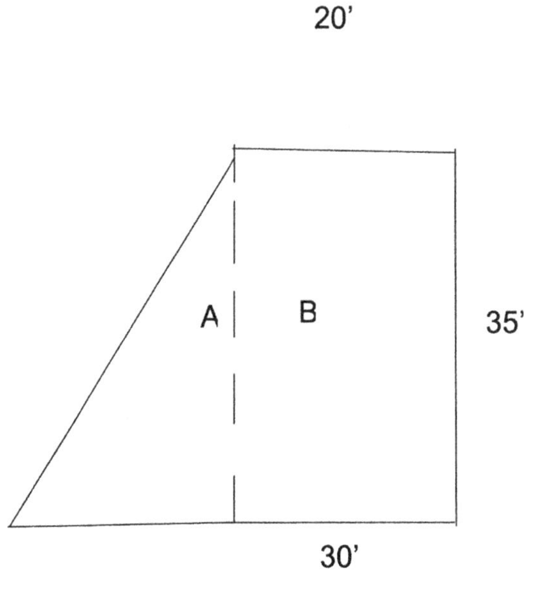

32. How many acres are in the diagram?

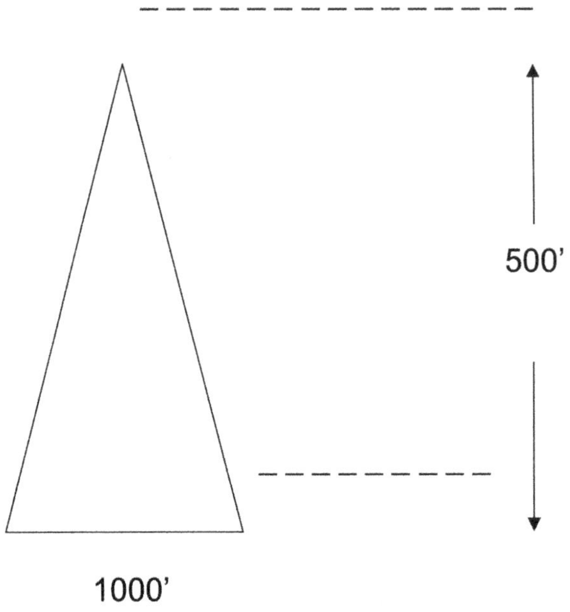

33. How many square yards are in the diagram above?

34. If you replace the pitched attic with a flat roof, how much space will be lost?

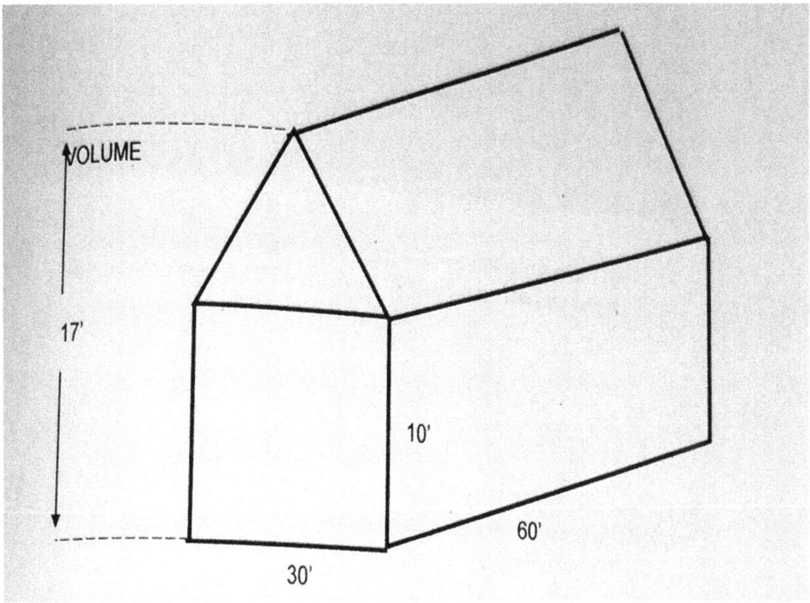
VOLUME
17'
10'
60'
30'

NOTES

Please feel free to write any notes in this section of the book.

AREAS, VOLUMES AND PYTHAGORAS

35. What is the length of side "C"?

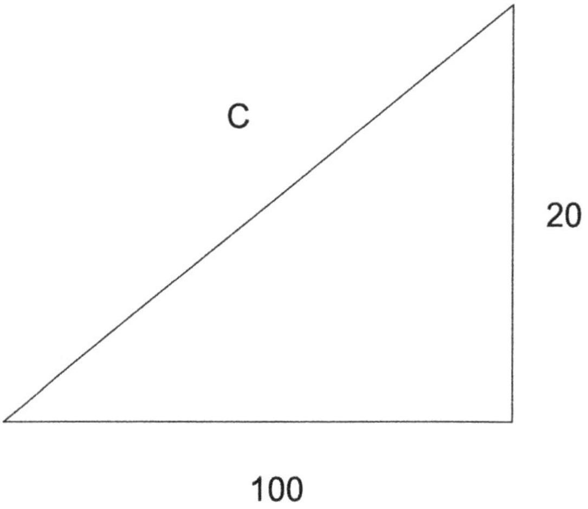

36. What is the volume of the building in cubic feet?

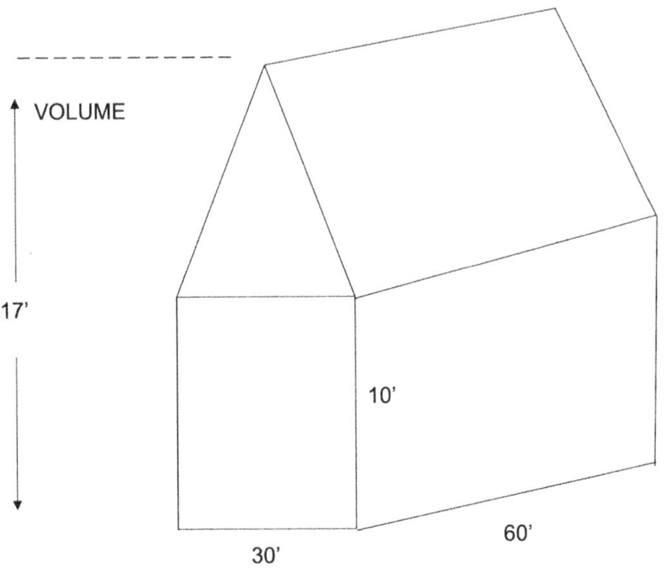

37. You want to add an above-ground pool to your backyard. To do so, you must remove part of the lawn. How many cubic yards will you dig out if the hole is 18 inches deep, 14 feet wide, and 21 feet long?

A. 16.3

B. 15.1

C. 17.2

D. 14.7

NOTES

Please feel free to write any notes in this section of the book.

IRV

Where I = annual net income

R = Capitalization Rate

V = Value

Use Part % and Total

38. A building valued at $85,000 produces a monthly income of $750. Maintenance expense amounts to $2,500 annually. What is the cap rate?

39. Someone purchased a piece of investment property for $250,000. The rent from 4 of 8 apartments is $500 per month; the rest of the apartments pay $580 per month. With the monthly expenses of $3,000, what is the cap rate?

40. A small professional office building produces a monthly net income of $2,750 from its annual gross income of $35,300. The cap rate derived from the market is 12%. What is the value of this property?

41. A property is earning a 9% rate of return with a value of $175,000. If a return of 11% is desired, what will the value of the building be?

42. John offers to sell an income property to David. The income is \$300 per month, and the total annual expenses are \$900. David figures on a 9% return on his investment. How much should David offer John for the property?

NOTES

Please feel free to write any notes in this section of the book.

MULTIPLIERS

GROSS RENT MULTIPLIER

GRM X GROSS RENT = VALUE

43. An appraiser, using the income approach to value, is estimating the value of a single-family residence. He finds comparable properties that rent for $750 per month. Recent sales were $85,000. What is the GRM?

44. The monthly income from a duplex is $700 per unit. If the GRM is 113, what is the value of the duplex?

NOTES

Please feel free to write any notes in this section of the book.

THE COST APPROACH

V = Cost - Depreciation + Land Value

45. What is the value of a brick bungalow 25' x 45' with a replacement cost of $50.00 per square foot? Lots in the area are selling for $20,000. The depreciation is estimated to be 15%.

46. John is appraising a two-story home 25' x 30' with a replacement cost of $40 per square foot. Depreciation is estimated to be 35%, and the land value is $22,000. What is the indicated value?

INTEREST

47. A loan was granted for 92% of an appraised value of $90,000. If the borrower's payment is $950 per month, and the annual interest rate is 10%, what would the third month's interest be?

48. A loan with a principal balance of $75,000 has 25 years remaining on the loan term. Monthly payments are $658.17. With an annual interest rate of 10%, what would the second month's interest be?

NOTES

Please feel free to write any notes in this section of the book.

POINTS

49. Steve obtained a loan from the local lender for 90% of the $67,800 sale price. The fee paid to the lender was $2,700. How many points does his fee represent?

50. If the service fee of $2,685 represents a 2% fee on an 80% loan, what is the property's purchase price?

51. The rate on the face of a note is 11.75%. The lender wants to increase the yield on the house note to 13.25%. The seller agrees to pay 40% of the total discount points due at the closing. How much of the points will the buyer owe on a $53,000 note?

PRORATION PROBLEMS

52. On July 15, the property on 123 Elm Ave will close. Real estate taxes for the year are $1,350.00, payable in arrears. Using the Statutory Month Method, how much are the prorated taxes?

53. In the example above, the seller prepaid the insurance premium of $350. How much was this prorated expense?

NOTES

Please feel free to write any notes in this section of the book.

EXTRA CREDIT

54. Phil's home is worth $57,000; his personal property is worth $12,500. If he wants to insure each for 80% of their market value and the insurance rate is $3.75 per $1,000 on the real property, with $5.50 per $1,000 on the personal property, what will the first year's premium be?

55. A property's value is determined to be $75,000. The property tax is based on 30% of the value. The district tax rate is 14 mills. What is the annual levy?

56. Ralph has just received his tax bill. He claims it is too high and wants to protest it. An appraisal indicated the home is valued at $63,900; however, Ralph would like $74,500 when he sells it. The county has been using 26% of the market value in determining the assessed value. The tax rate is $3.95 per 100 of assessed valuation. An equalization factor in the neighboring county has been at 1.15. None is used in Ralph's county. How much is Ralph paying?

57. Alice purchased a home for $78,035. How much was the transfer tax paid to the state?

58. A house sold for $125,000. The sponsored licensee received $2,625 as their share of the commission. If the total commission paid to the company was 7%, what was the sponsored licensee's percentage share?

59. Pete sold a home for $186,000, and the managing broker retains 50% of the six-percent commission for the company. The listing licensee receives 20% of the balance. How much will be paid to selling broker Pete? The commission can also be set in a graduated percentage schedule.

60. A sponsoring broker gets 6% of the first $100,000 and 3% of any amount over $100,000. What would be the sponsoring broker's loss if a house listed for $180,000 has to be reduced by 20% for sale?

61. Marty entered a listing contract with the broker agreeing to pay the sponsoring broker a 6% commission upon the sale of the property. Marty has informed the sponsoring broker that he must net $65,000 to buy a new home. The sponsored licensee working for the sponsoring broker is entitled to 53% of the total commission. How much will this sponsored licensee receive if the home sells for the listed price?

62. Joe bought a home for $75,000, built a swimming pool costing $4,500, and spent $625 on filtering and cleaning equipment. The following month he received an unexpected promotion and had to relocate to another state. If he listed his property at $85,000, how much would he expect to make? What would be his percentage of profit?

63. What would be the book value of a building three years from now? Assume the building has depreciated 3% each year, is 27 feet wide by 43 feet deep, and cost $50 per square foot of the building? The estimated land value is $10,000.

64. A tract of land is 90 feet by 236 feet. You own one-half of the tract; your brother owns 30% of the balance. What is the square footage of the remaining portion?

65. The land is selling for $6.08 per square footage. Estimate the value of a parcel of land with front footage of 64 feet and a depth of 205 feet.

66. A purchaser has insulated his dwelling by covering the attic floor 24 feet by 32 feet of rock wool. Each bag of rock wool will cover 36 square feet. How many bags will be required?

67. Tom wants to buy a business with a monthly expense of $950, representing 75% of the business's gross monthly income. If Tom wants a 13% return, what should he be willing to pay for the business?

68. A property showed a net income of $3,500 per year over the past five years. Assuming the property will yield a return of approximately 8%, what is the value of this property?

69. Paul owns a brick and frame 2-story dwelling built-in 1920. The reproduction cost is $250,000, and the effective age is 20 years, with an economic life of a hundred years, the land value is estimated at $75,000. Using the cost approach, what is the value of Paul's property?

70. Using the information in problem 69, Paul rents the property for $2,000/month. There are $500/month in expenses and a multiplier derived from the marketplace of 137. Using the income approach, what is the value of Paul's property?

NOTES

Please feel free to write any notes in this section of the book.

MATH REVIEW ANSWERS

1. CHANGE THE FOLLOWING PERCENTAGES TO A DECIMAL.

12.5%	.125	34%	.34	5%	.05
125%	1.25	34.5%	.345	5.5%	.055
14%	.14	.25%	.0025	33%	.33
140%	1.4	.45%	.0045	25%	.25
9%	.09	.34%	.0034	7.5%	.075
19.5%	.195	.034%	.00034	.09%	.0009

2. CHANGE THE FOLLOWING DECIMALS TO A PERCENTAGE.

.15	15%	.02	2%	.29	29%
.025	2.5%	.257	25.7%	.357	35.7%
.0976	9.76%	.0078	.78%	.154	15.4%
.2	20%	.26	26%	.788	78.8%

3. CHANGE THE FOLLOWING FRACTIONS TO DECIMALS.

$\frac{7}{8}$ =.875 $\frac{15}{16}$=.938 $\frac{5}{8}$=.625 $\frac{3}{4}$=.75

4. CHANGE THE FOLLOWING PERCENTAGES TO A DECIMAL.

$7\frac{7}{8}$%=.07875 $9\frac{15}{16}$%=.09938 $14\frac{5}{8}$%=.14625 $7\frac{3}{4}$%=.0775

5. 8,000/.08 =100,000

6. 14,157/43,560 = .325 OR 32.5%

7. 1/3 = .3333 -> 5/16 = .3125 -> .3333 -> 1.0000

$$\begin{array}{r} +\ .3125 \\ \hline .6458 \end{array} \qquad \begin{array}{l} -\ .6458 \\ \hline .3542 \text{ or } 35.4\% \text{ (C)} \end{array}$$

8. 657.49/.085 = 7735 (C)

9. 85250 X .07 = 5967.50 5967.50 X .125 = $745.93

10. 78.375 X .98 = 76,807.50 76,807.50 X .0725 = 5568.54

11. 55,900

x .0675

3773.25

x .40

1509.30

12. 57/500 .114 or 11.4% (B)

13. 100% 140

- 75% 25% = .25 x .25

25% 35

14. 140,000 x .80 = 112,000 insurable value

Insurance rate = 3.68 /1000 = .00368

112,000 x .00368 = $412.16

15. $ 106,500 [Tax Rate = 2.50 = .025]

x .25 100

$ 26,625

x 1.35

$ 35,943.75

x .025

$ 898.59

16. $ 83,625 Tax Rate: 3.78 100 = .0378

x .0378

$ 3,161.02

17. One mill = 1/100 = .001

Tax Rate = 45 mills = 45/1000= .045

$75,000

x .40

30,000

x .045

$ 1,350

18. If you multiply going <u>**DOWN**</u> the equation, you must divide when going <u>UP</u>:

The tax rate = 25/100 = .025

242.50 .025 = $9,700 (Assessed Value)

$9,700 20 = $48,500 (Value)

19. 49,500/500 = 99

99

x .50

$49.50

20. 63,495

- 22,875 (assumed mortgage)

40,620 (round to)

41,000

500

82

x .50

$41.00

21. 75,000 = 80,645

.93

22. 71,100 = 75,638.29

.94

23. 58,900 -> 61,400 = 65,668

+ 2,500 .936

61,400

24. 4000 -> 3000 -> 1000 = .125 or 12.5%

x 2 x 3 8000

8000 9000

25. 125,000 = 100,000

1.25

22. 125,000 = $166,666

.75

23. D

24. 425,000 = 10,625 10,625 425,000 297,500
40 x 12 -127,500 + 85,000
127,500 297,500 382,500

25. ½ of ½ of ¼ or 2 x 2 x 4 = 16
640 16 = 40 acres

26. A - ½ x ¼ = 1/8 x 640 acres = 80 acres
B - ¼ x ¼ = 1/16 x 640 acres = 40 acres
Total area = 120 acres

27. L x W = AREA
26 x 34 = 884 SF
884 9 = 98.2 Sq Yds

28.

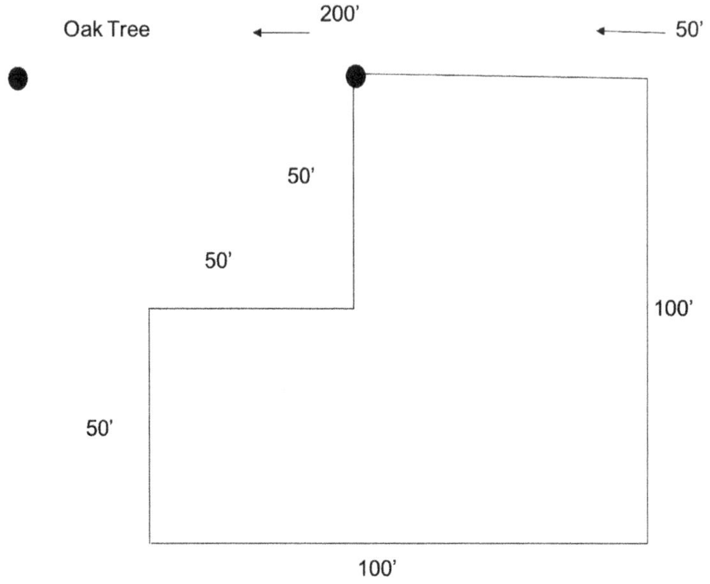

[Not To Scale]

A. 50’ x 100’ = 5,000 SF

50 x 50 = <u>2,500</u> SF

7,500 SF

B. 7,500 SF @ $5/SF = $37,500

C. 7,500 43,560 = .172 = 17.2%

D. 50 + 100 + 100 + 50 + 50 + 50 = 400 linear feet

E. $6 x 400 feet = $2,400

29.

205 13,120

<u>x 64</u> <u>x 6.08</u>

13,120 79,769.60

30.

24
x 32 -> 768 / 36 = 21.33, or 22 whole bags.
768

31. A: (B x H) 2 = AREA
(35 x 10) 2 = 175 SF
B: (L x W) = AREA
(20 x 35) = 700
700 + 175 = 875 SF (Total)

32. (B x H) 2 = AREA
(500 x 1000) 2 = 250,000
250,000 43,560 = 5.73 acres

33. 250,000 9 = 27,777 Sq Yds

34. You Lose The Whole Attic (Triangular Space) Which Was 6,300 Cubic Feet

35.

$$a^2 + b^2 = c^2 \text{ or } \sqrt{a^2 + b^2} = c$$

Therefore: $\sqrt{100^2 + 200^2} = c4$

$\sqrt{10{,}000 + 40{,}000} = c$

$\sqrt{50{,}000} = c$

223.6' = c

36. L x W x H = Cubic Feet of Cube (B x H x L) 2 = Cubic Feet Triangle
60 x 30 x 10 = 18,000 CF
(30 x 7 x 60) 2 = 6300 CF
Total: 18,000 + 6300 = 24,300 CF

37. Change 18 inches to a decimal: 18" = 1.5 Feet

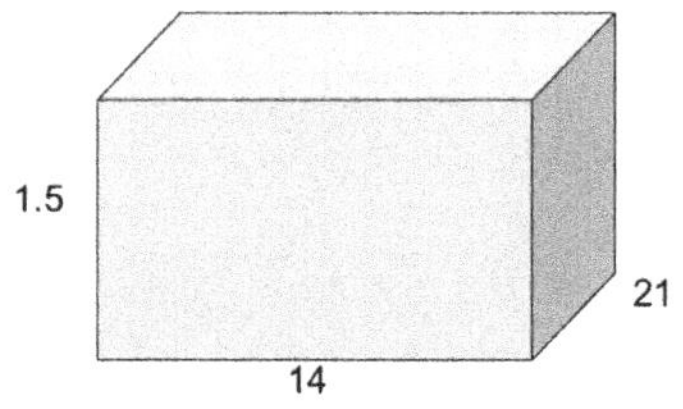

L x W x H = Volume

21 x 14 x 1.5 = 441 Cubic Feet

411 27 = 16.33

# of Cubic Feet in a Cubic Yard

38. 750 -> 9000 -> 6500 = .076, or 7.6%

x 12 - 2500 85,000

9000 6500

39. 500 580 3,000 24,000 15,840 =.063

x 4 x 4 x 12 +27,840 250,000

2000 2320 36,000 51,840

x 12 x 12 -36,000

24,000 27,840 15,840

40. 2,750 -> 33,000 = 275,000

x 12 .12

33,000

41. 175,000 x .09 = 15,750 -> 15,750 = 143,181

.11

42. 300 3600 2700 = 30,000

x 12 - 900 .09

3,600 2,700

43. 85,000 = 113.33

750

44. 700 1,400

x 2 x 113

1,400 158,200

45. 25 -> 1125 -> 56,250 -> 56,250.00 -> 47,812.50

x 45 x 50.00 x .15 - 8,437.50 + 20,000.00

1125 sq. ft. 56,250.00 8,437.50 47,812.50 67,812.50

46. 25 -> 1500 -> 60,000 -> 60,000 -> 39,000

x 30 x 40.00 x .35 - 21,000 + 22,000

750 sq. ft. 60,000 21,000 39,000 61,000

x 2 stories

1500

47. 90,000 -> 82,800 loan -> 950 PMT -> 82,800 loan Bal. After 1st mon.

x .92 x .10 - 690 i - 260 i

82,800 8280 i for 1 year 260 Prin. paid 82,540 +12

690 i for 1 month

Second Month

82,540 -> 950.00 PMT -> 82,540

x .10 - 687.83 i - 262.17

8254/12 262.17 Prin. Pd, 2nd month 82,277.83 Bal. After 2nd mon.

687.83 i for 2nd month

Third Month

82,277.83 -> 950.00 PMT 82,277.83

x .10 - 685.64 i - 264.36

82,227.78/12 264.36 Prin. After 3rd month 82,013.47 Balance

685.64 i for 3rd month = **Answer to problem**

48. 75,000 -> 658.17 -> 75,000.00 -> 74,966.83

x .10 - 625.00 - 33.17 x .10

7500/12 = 625 int. 33.17 74,966.83 7,496.68/12= 624.72 int. 2nd mon.

49. 67,800 -> 2700 = .0442

x .90 61,020

61,020

50. 2685 = $134,250 $134,250 = $167,812 (rounded)

.02 .08

51. .1325 -> 53,000 -> 6360

- .1175 x .12 x .60 If the seller pays .40 of points,

.015 or 1.5% 6,360 3816 then the buyer would pay .60

1/8 = 1 point

So, we have 12 points or 12%.

52. $1350 / 12 = $112.50/month

112.50 / 30 = $3.75/day

112.50 x 6 = $675.00

and

3.75 x 15 = 56.25

$731.25

Therefore, there is a $731.25 debit to the seller, and credit to the buyer.

53. 350 / 12 = 29.17/month

29.17 / 30 = .97/day

29.17 x 5 = $145.85

and

.97 x 15 = 14.55

$160.40

Therefore, there is a $160.40 debit to the buyer and credit to the seller.

54. (A) (B)

57,000 x .80 = 45,600 12,500 x .80 = 10,000

45,600 / 1000 = 45.6 10,000 / 1000 = 10
45.6 x 3.75 = $171.00 10 x 5.50 = $55.00
Total: $171 + $55 = $226

55. 75,000 **56.** 63,900 **57.** 78,035 Round up to:
x .30 x .26 78,500/500
22,500 16,614 /100 =157
x .014 = 166.14 x.50
$315 x 3.95 78.50
656.25

58. 125,000 -> 2625 = .30
x .07 8750
8750

59. 186,000 -> 11,160 -> 5580 -> 5580
x .06 x .50 x .20 x .80
11,160 5580 1116 4464

60. 180,000 36,000
x .20 x .03
36,000 1,080

61. 65,000 = 69,148.93 -> 69,148.93 -> 4148.93
.94 x .06 x .53
4,148.93 2,198.94

62. 75,000 -> 85,000 -> 4875 = .061
4,500 -80,125 80,125
+ 625 4,875
80,125

63. 27 -> 1161 -> 58,050 -> 1741.5 -> 58,050.00 -> 52,825
x 43 x 50.00 x .03 x 3 - 5,224.50 + 10,000
1161 58,050 1741.50 (*) 5224.5 52,825.50 62,825

(*) = Depreciation Per Year

64. 236-> 21,240 -> 10,620 -> 10,620

x 90 x .50 x .30 - 3,186

21,240 10,620 3186 7,434

Your Share Brother's Share Remaining Balance

65.

205 -> 13,120

x 64 x 6.08

13,120 79,769.60

66.

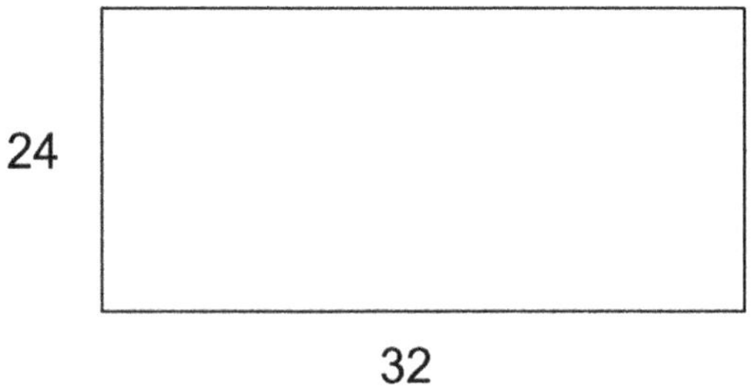

24 * 32 = 768 -> 768/ 36 = 21.33 = 22 bags

67. 950 11,400/.75 = 15,200 15,200 3,800/.13 = $29,230.77

x 12 -11,400

11,400 3,800

68. 3,500 = 43,750

.08

69. 250,000 - 50,000 = 200,000 (20/100 = .20 .20 x 250,000 = 50,000)

200,000 + 75,000 = 275,000

70. 2,000 x 137 = 274,000. (Note: When using a multiplier, we don't analyze expenses. Also, the value from each approach does not need to match. The appraiser uses the reconciliation process to render a final opinion of value.)

ABOUT SAM MARTIN, PHD, DREI, MOT

Meet Sam Martin, PhD, DREI, MOT

Sam Martin PhD, DREI, ASA, CDEI, MOT is the owner of SamTheTutor.com, LLC. Dr. Sam has a PhD in Education and is one of 500+ Certified Uniform Standards of Professional Practice (USPAP) instructors worldwide. He is a Distinguished Real Estate Instructor (DREI), a prestigious designation granted by the Real Estate Educators Association (REEA). He holds North Carolina Certified Residential Real Estate Appraisal, Real Estate Broker, and real estate instructor licenses. Dr. Sam has been teaching appraisal and real estate classes since 1988. He has taught for a variety of institutions, most recently: Superior School of Real Estate; Real Estate Express; and The Chicago Association of Realtors. Dr. Sam has written an advanced appraisal text book used in certification courses. Dr. Sam develops and teaches valuation courses for Realtor University, an online graduate-level institution where he is a Professor Emeritus.

Dr. Sam has taken advanced educational technology courses from the Illinois Online Network (ION), where he received the Master Online Teacher (MOT) certificate. Dr. Sam holds the Certified Distance Education Instructor (CDEI) designation from IDECC (www.idecc.org).

The Real Estate Education Foundation (REEF) has awarded Dr. Sam the Academic Scholarship three times. Dr. Sam has won three grants from the Association of Illinois Real Estate Educators (AIREE). Dr Sam was named the Educator of the Year in 2014 by AIREE. The Chicago Association of Realtors Educational Foundation has awarded Dr. Sam a Higher Education Scholarship three times.

Visit Sam at www.samthetutor.com

ABOUT SUE MIRANDA-ROSENTEEL

Meet Sue Miranda-Rosensteel

Sue Miranda-Rosensteel has specialized in real estate education for over twenty years. She was the founder and President of Miranda Real Estate School in Chicago, a company she grew into one of the largest privately-owned schools in Illinois. Sue has served as Director of Real Estate Education, Senior Manager of Real Estate Curriculum, and National Real Estate Expert for other top leading schools. In addition to be the President and owner of One-On-One Real Estate Center, she is also an owner partner of SPIRE Real Estate Education, a company specializing in professional training, career development, pre-licensing, post, and continuing education courses. She is licensed to teach pre-licensing and continuing education classes in Illinois, Indiana, and Pennsylvania.

Sue says that Real Estate Education is her passion and looks forward to being part of the lead team of instructors offering live classes for SPIRE. She enjoys all facets of the field – from working one-on-one with students and instructors - to writing and creating courses that serve a purpose. She currently volunteers her time to serve as a Director for the Real Estate Educators Association (REEA), and engages in various committees for ARELLO and AIREE. She strongly believes it's her time to give back to the community that supported her growth throughout the years and help build the next generation of real estate licensees and educators.

Made in the USA
Las Vegas, NV
22 March 2022

46027879R00066